NewScientist

The Book of 100 Things to Do Before You Die

Edited by Valerie Jamieson and Liz Else of New Scientist

Introduced by Bob McDonald, host of CBC Radio's Quirks & Quarks

First published in Great Britain in 2004 by Profile Books Ltd.
Hardcover edition published in 2006 by House of Anansi Press Inc.

This edition published in 2007 by
House of Anansi Press Inc.
110 Spadina Avenue, Suite 801
Toronto, ON, M5V 2K4
Tel. 416-363-4343
Fax 416-363-1017
www.anansi.ca

Distributed in Canada by
HarperCollins Canada Ltd.
1995 Markham Road
Scarborough, ON, M1B 5M8
Toll free tel. 1-800-387-0117

House of Anansi Press is committed to protecting our natural environment. As part of our efforts, this book is printed on Rolland Enviro paper: it contains 100% post-consumer recycled fibres, is acid-free, and is processed chlorine-free.

11 10 09 08 07   1 2 3 4 5

LIBRARY AND ARCHIVES CANADA CATALOGUING IN PUBLICATION DATA

The book of 100 things to do before you die / edited by Valerie Jamieson and Liz Else of New Scientist.

Previous title: Have your poo rolled away by dung beetles and 99 other things to do before you die, 2006.
ISBN-13: 978-0-88784-784-4
ISBN-10: 0-88784-784-6

1. Science—Popular works. I. Jamieson, Valerie II. Else, Liz III. Title: The book of one hundred things to do before you die.

Q164.B664 2007   500   C2007-902662-1

Cover design: Bill Douglas at The Bang
Text design and typesetting: Tannice Goddard

Canada Council for the Arts   Conseil des Arts du Canada

We acknowledge for their financial support of our publishing program the Canada Council for the Arts, the Ontario Arts Council, and the Government of Canada through the Book Publishing Industry Development Program (BPIDP).

Printed and bound in Canada

Introduction

Stepping out of a perfectly good airplane at 3400 metres without a parachute is not natural. Yet there I was, plummeting straight down at 250 kilometres an hour in the most thrilling experience of my life. The reason I'm here to tell about it is because there was a guy with a parachute built for two strapped to my back and he made sure that our landing was a soft one. Life-affirming experiences like tandem skydiving or climbing Mount Everest may not be your cup of tea, but you don't have to put your life on a limb to try a novel activity that opens your eyes to the wonders of the world around you. Simple things, like, well, having your poo rolled by a dung beetle, are fascinating insights into other worlds within our own, where insects see our waste as a resource, or where the ground beneath your feet is new skin for the Earth, freshly spewed from the mouth of a volcano.

In every case, the unique experience contains an additional surprise. You might think that skydiving feels like falling. It doesn't. The overwhelming sensation is, in fact, one of being pushed upwards. Once you step into thin air, you quickly discover that it's not so thin after all. As your body accelerates downwards, the force of the air against your chest becomes stronger. After just a few seconds, that upward force of the air is the same as the downward pull of gravity and you reach a steady speed called terminal velocity. It feels like floating, not falling.

From more than two kilometres up, the ground is distant and firm, while you remain suspended in a very big, very soft blue space. For 30 long seconds, as the wind pulls your cheeks back towards your ears in a goofy grin, you wave at your ungainly flying companions, all defying natural law, all on the edge of instant annihilation.

As a science journalist I'm fortunate to follow the cutting edge of human knowledge as humanity discovers the complex, often silly aspects of life. I'm also fortunate to have travelled a great deal to see the world for myself, one of the best educations available.

While drinking Ouzo at the base of the Parthenon in Greece a fellow traveller once asked, "What do you do for a living?" I replied, "Right now, I'm fulfilling childhood fantasies." Which was exactly true. At that time I was part way through a six month solo journey around the world, the best thing I've ever done for myself. With no plans and nothing more than a backpack, I followed my dream to circle the globe before I die. Along the way I touched the pyramids, gazed up at Mt Everest, saw thundering herds of wildebeest on the Serengeti plains of Africa and saw a total eclipse of the sun.

We live in remarkable times, when humanity uses the tools of science to discover the world in exquisite detail. The last 500 years have been spent discovering planet Earth on ships, planes and rockets, finally providing a distant glimpse of this beautiful blue ball we live on. Now, through the most powerful tools ever wielded, we've driven on other worlds in space, manipulated the structure of DNA and travelled in time through fossilized bones that tell tales of strange creatures, including our own ancestors, who lived at a time when Earth was a very different place.

When you think about the world through our five senses, it's really quite small. We can't see very far, and from our perspective on the ground, the Earth looks pretty flat. For every mountain there's a valley and if you walk far enough in any direction you will eventually come to an ocean. No wonder people a thousand years ago believed the world to be an island in the middle of a circular sea. That's the way it looks to our eyes! But through the eyes of science this island world becomes a tiny speck floating through an unimaginably large and beautifully complex universe, while another microscopic universe thrives in a drop of water.

The activities in this book are opportunities to put on scientific spectacles and see details in your everyday world that have been hidden from view. Doing these activities can even be good for you. Experiments on the brains of people who died of Alzheimer's disease have clearly shown that those who remained mentally active into old age had less deterioration of brain cells. Seniors who learned to play a musical instrument, learned a new language or even solved crossword puzzles developed the disease later than their less active counter parts. And when they did get it, the disease showed up later in life and developed more slowly. The brain is a use it or lose it organ. Every time we challenge our minds with a new task or experience something novel the brain creates new neural connections to preserve that experience. In the same way that muscles grow stronger with exercise, the brain grows more capable with new experiences.

How often have you said, "One day I'm going to . . ." Well, do it. And if you can't think of something interesting to do, this book is full of great suggestions. So have a look at your own DNA, see the rings of Saturn and, yes, have your poo rolled away by a dung beetle. The world will never be the same.

— *Bob McDonald*

Swim in a bioluminescent bay

A handful of pixie dust

Want to make shimmering snow angels in water? Or find out how it feels to be a human comet with a bright tail? Then take a trip to a bioluminescent bay. There are only a few in the world, but one of the brightest and most pristine is Mosquito Bay in the small Caribbean island of Vieques, off the coast of Puerto Rico.

Pick a moonless night and use a kayak if you can. Each stroke will light up the water and your boat will be surrounded by the most magnificent bright blue-green glow.

Dip your hand in the water and twirl it around — the more vigorous your movements, the more you'll shine. Cup your hands and reach in. As you lift, look closely at the water. There you'll see the culprits: microscopic organisms called dinoflagellates that release light when they are disturbed.

Nature's pixie dust is all around you.

Be a human particle detector

I spy a cosmic ray

En route to the moon, the Apollo astronauts all reported seeing mysterious flashes of light. They were caused by cosmic rays, super-fast atomic particles, hurtling through their eyeballs. These rays travel within a whisker of the speed of light and emit the visual equivalent of a sonic boom as they smash their way through fluid in the eye. Known as Cherenkov light, this eerie blue glow is detected by the retina.

You can't have the same experience on Earth because the planet's magnetic field protects us from the brunt of cosmic radiation. But you don't have to travel as far as the moon to see a firework display going off in your eye.

SpaceShipOne recently became the first privately built rocket plane to reach space, and pundits are predicting plenty more flights will follow. Soon you may be able to hitch a ride to the edge of space and see cosmic nuclei zipping through your eyeball. Then you'll really know what it feels like to be a human particle detector.

Patrick Moore

Discover a comet

I have always been an observer of the moon, but I have never scoured the sky for comets. Before I die, I would love to discover a bright comet as magnificent as the one Italian astronomer Giovanni Donati saw in 1858. It would be great to see Moore's Comet blazing across the sky.

Patrick Moore presents BBC Television's *The Sky at Night*

Stand at the South Pole

90 degrees south

It's a once-in-a-lifetime trip so it won't be cheap. Flying to the South Pole is pretty expensive, but flying is still better than the old method of trudging there on foot, dragging your sledge behind you.

Once you've arrived, look for the ceremonial striped barber's pole where all the dignitaries have their photos taken, before heading for the true pole, where all the Earth's lines of longitude meet at your feet. This is marked by a more modest thin metal pole bearing a brass cap, which is shifted by around 10 metres each year to account for the sliding glacier beneath.

Beware of altitude sickness: the ice is so thick at the South Pole that its surface is more than 3000 metres above sea level. And it lies so far from the sea that the air is as dry as the Sahara and bitingly cold, even in the height of summer. You may still experience a blizzard there if the winds pick up the surface coating of snow.

Go in the summer to experience a midnight when the sun never sets, but bounces eerily around the horizon for a full 24 hours. And watch out too for airborne crystals of ice known as diamond dust that can cast an eerie ring of light around the sun.

Have your poo rolled away by dung beetles

You can't get closer to nature

If you ever find yourself caught short out in the wild, don't despair: a humiliating situation could turn out to be your best way of getting up close and personal with nature.

Once you've done your thing, watch and wait. Dung beetles have a keen sense of smell and competition for poo is fierce, so it won't take long for them to appear. They will scuttle boldly up to your deposit and begin rolling balls of it away, head-butting it and pushing it with their forelegs. They will vanish with their prize as quickly as they appeared. Once safe, your dung may be used for food and as a breeding ground.

The best place to experience this is in the desert. Here the dung beetles are very large, very hungry and very easy to see against the golden sand. But you don't have to venture that far. There are over 7000 species of dung beetle and they live in every continent except Antarctica.

Not only will the experience remind you that we are still animals, it will show you that we are part of a delicate ecosystem that uses everything. What we regard as waste is like a lottery jackpot to another species.

Search for signs of extraterrestrial intelligence

Just pick up a phone, ET!

Be the first human to make contact with an alien and you'll go down in history. Of course, some people already claim to have been abducted by aliens. But no one really takes them seriously.

The best chance you have of being believed is if your home computer spots ET's calling card. Simply download the SETI@home program from setiathome.ssl.berkeley.edu and your computer will search through candidate signals picked up by the Arecibo radio telescope in Puerto Rico. Because it's a screen saver, the program runs while you're having lunch or working out.

If your computer sees an interesting signal, you'll get full credit as it is investigated by the SETI@home team and dozens of other groups worldwide. And if it does turn out to be ET, you'll be remembered for the greatest astronomical discovery of all time.

Go down a deep mine

Journey towards the centre of the Earth

Descending into a mine is the closest you're going to get to the Earth's core. The deepest mine of all is South Africa's aptly named Western Deep mine, west of Johannesburg, which goes down for a staggering 3.5 kilometres.

Wearing your crash helmet and regulation jumpsuit, you travel downwards in a series of lift cages — nobody has made a cable long enough to make this a non-stop trip.

At this extreme depth the walls are hot to touch and the air would be an unbearable 60°C if it weren't for the air-cooling systems and water sprays. Even then, the temperature rarely drops below 35°C and the humidity is close to 100 per cent. You'll still sweat so much that your boots will slosh as you walk.

As you contemplate the crushing pressures that the walls of the tunnel are under, marvel that ultra-resilient microbes are found living even here. And then realise that this journey is still only a tiny scratch on the Earth's surface. There are another 6375 kilometres to go to the centre of the Earth.

Home Lab: Extract your own DNA

See what makes you unique

Want to know what you're really made of? How better than to take a look at your DNA. You don't need fancy machines to do this, just some salt, a strong drink and some washing-up liquid.

First, get a sample of your own cells by swilling salty water vigorously round your mouth for half a minute. Spit it out into a small glass containing a teaspoon of washing-up liquid diluted with three teaspoons of water.

Give this a good stir for a few minutes so that the detergent can break open the cells and release the DNA.

Then, very gently, pour a couple of teaspoons of ice-cold strong alcohol down the side of the glass. Gin works fine. If you're careful, it will form a separate layer on top of the spit-salt mix. Soon you'll see spindly white clumps forming in the alcohol. This is your DNA.

Somehow you can't help expecting your DNA to wriggle or have flashing lights. But there it is — this white goo is what makes you unique.

See Halley's comet . . . twice

Club 18 to 40

Halley's comet is without doubt the most famous giant snowball in space. It is named after the English scientist Edmond Halley who predicted that the comet, which he saw in 1682, would swoop round the solar system and return in 1758. And sure enough it did. Today the comet still appears in the night sky about every 76 years as it nears the sun and starts to boil up in the heat.

Comet Halley disappointed many people on its last flyby in 1986 when it was barely visible to the naked eye. But in 1066, the comet was so shockingly bright it terrified millions across what is now Europe. Some say it cemented the Norman victory at the Battle of Hastings, making the Saxon king Harold of England and his followers feel doomed to failure.

Seeing Halley's comet twice must be an incredible experience, but it's not possible for everyone. To join the double-Halley club, you must have seen the 1986 visit, making you at least 18 years old today. And you'll have to live to see the next visit in 2061, which means you are probably now less than 40 years old.

If that rules you out, don't despair. You're in the good company of Halley himself, who never lived to see the return of the comet that made him famous.

Deliver a baby animal

Welcome to the world!

This is one of life's most surprising and moving experiences. And pretty accessible. Farmers are often only too happy to have extra help during lambing and calving. For a more exotic experience, ask a zoo keeper if you can be involved with the birth of a zebra, camel or giraffe.

If you're helping a cow, make yourself useful when her waters break and you see the membrane of the water sac hanging out. Strip, soap and disinfect your arm up to your shoulder, slide it inside her, and feel around. You can check how dilated she is, how often she's contracting, and whether you can feel the feet and nose of the calf.

Two to six hours after her waters break, the cow will lie down and start contracting very regularly. Within an hour you should see the front legs of the calf poking out, and then the nose.

Eventually the strong smell of disinfectant and soap will be overpowered by pungent birth fluids. They will get the mother's attention too. She'll look round and deliver the rest of the calf a few minutes later. She may well get up quickly, turn around and start licking the newborn. You can help too, by cleaning mucus off its nose and checking its breathing.

And for a moment, you'll feel as proud as she does.

Be a gecko

Cling on in there

Just for a day you too could be . . . a gecko. Just think: you could walk up walls, scuttle to the top of tall buildings, or lurk on the ceiling waiting to give your friends a shock. Best of all, you can find out how it feels to do something distinctly non-human.

Researchers at the University of Manchester in the UK and Carnegie Mellon University in Pittsburgh are developing synthetic materials that mimic the ceiling-clinging power of the gecko. Like a gecko's toes, these materials are covered with millions of tiny hairs that will cling to almost any surface thanks to weak intermolecular forces.

Already the Pittsburgh group has built a robot gecko that can climb walls using this technology. At the moment, the materials are too primitive to make gecko gloves that will safely hold a human's weight, but the researchers are working on new versions that should be much stronger.

So before very long you might be able to buy a pair of gecko gloves — and perhaps some similarly sticky socks and kneepads — to let you enjoy a lizard's freedom from gravity. Off you scuttle then . . .

Visit Hiroshima

Humankind's darkest hour

Outside the Hiroshima Peace Memorial Museum, you'll pass the shattered dome of the one building left standing close to where the atomic bomb exploded on 6 August 1945. You'll read the statistics. Within half a kilometre of this spot, almost everyone died within a month. Around 140,000 people died by the end of the year. But it is inside the museum that death becomes more personal.

In the cool dim interior you'll see the shadow of a man, sitting on the steps of a bank waiting for it to open, that is burnt into the stone. The man vaporised. Then there are the tattered, burnt clothes of the schoolchildren who died within the day.

Most moving of all is Shin-ichi's tricycle, sitting in a glass case, burnt and rusted. Shin-ichi was three years old. He was outside playing with his tricycle when the bomb dropped. His brother and sister were burnt to death in the first flash. He survived until the night. His parents had been inside the house and survived. Not wanting to consign Shin-ichi to a lonely graveyard, they buried him in their own backyard along with the tricycle. Forty years later, his father dug it up and gave it to the museum.

When you emerge into the sunlight again, expect to be a changed person.

Save a species

It's alive!

Make your next holiday one to remember by saving a species from extinction. All you need is the right destination, a pair of binoculars, a camera and lots of luck.

For your best chance, catch sight of a bird so rare that scientists have given up hope that it still exists. Try a walk through the pine forests of the Sierra Madre Occidental mountains in north-west Mexico. The forest was once home to the glorious imperial woodpecker (*Campephilus imperialis*), the largest woodpecker in the world. The last confirmed sighting was in 1956, although locals claim to have seen the bird eight times since.

If you spot a bird about 60 centimetres high with black plumage, white flashes on its wings and sporting a stunning red crest, take a picture and send it to BirdLife International. You'll spark a rescue effort that could bring the species back from the brink.

It has happened before. In 2003, the long-legged warbler (*Trichocichla rufa*) was photographed on Fiji's largest island of Viti Levu. The bird had not been seen since 1894 and conservationists are now trying to protect it.

You'll feel good about your holiday for ever.

Listen to an iceberg being born

Thrills and chills

For anyone who has ever doubted the power of ice, one amazing encounter will set you straight. Some of the world's biggest, fastest glaciers shoulder their way directly down to the sea, or perhaps a milky glacial lake, and at their base you will find a sight like no other.

Argentina's Los Glaciares National Park in Patagonia is one of the most outstanding glacial areas of the world and the Perito Moreno glacier is a highlight. Rising tens of metres out of the slushy waters of Lake Argentino is a fractured cliff of impossibly blue ice, crowned with teetering spires. And it's moving. It may not look like it, but the pings, rifle cracks and terrible squeals of millions of tonnes of tortured ice soon tell you this monster is restless.

Although you know it's just a matter of time before something gives, nothing prepares you for the thunderous roar of an iceberg being born. In slow motion that suddenly betrays the scale of the thing, an ice block the size of a small building breaks off and explodes onto the water, and the din echoes from the mountains. Just be sure not to get shredded by ice shrapnel — it has happened to visitors who strayed too close.

Inhale helium

Sing like Mickey Mouse

The old ones are the best, and this is a classic everyone should try. Helium's low density means that sound waves travel through it nearly three times as fast as through air. This radically alters the resonant frequencies of your throat and makes you sound, frankly, ridiculous. So take a good lungful from a balloon and burst into a few bars of *Nessun Dorma* — if you can manage not to laugh as soon as you hear the first note.

A word of warning, though: to avoid asphyxiating yourself get plenty of fresh air between breaths of helium, and go easy on the stuff. Also, never try to breathe helium direct from a pressurised tank, or you probably won't be singing anything. Ever.

Swim with great white sharks

The ocean's most fearsome predator

Here's your chance to get up close to a supreme killing machine. One of the best places to see great white sharks is Gansbaii, near Cape Town in South Africa. Here only a limited number of cage-diving expeditions are permitted, in order to protect these endangered animals.

Go in winter, when great whites congregate in Shark's Alley between Dyer Island and Geyser Islet, feasting on fur seals. What will probably strike you is how cautiously the sharks approach the boat. Though they are fearsome killers, like any wild animal they are wary of the unknown. In spite of their reputation, they don't even enjoy human flesh.

When you climb into the cage underwater, the large size of the mesh might make you nervous. Until, that is, the great grey shape looms by in the water and you appreciate the scale of these animals — they're about the same size as a family car. Calm down: you have too little fat on your bones to be truly palatable to a great white. If you were unlucky enough to be bitten by one, it would spit you out — adding the ultimate insult to your injury.

Experience the nocturnal world

Turn off the light

Pick your favourite place: perhaps a rural beauty spot, a beach or even your own garden. Instead of visiting by day, go in the evening and stay up all night. Watch as it is transformed into a strange nocturnal world of new sights and sounds that you normally miss when tucked up in bed.

Empathise with a machine

From Dr Frankenstein's monster to *Star Trek's* Commander Data, there has been a long fascination with the idea of man building a man. Over the last 10 years or so, a few academics around the world have taken this seriously and built robot humanoids. So far these robots have been zombies — they have the outward physical shape of a human, they make eye contact and they sometimes smile and respond in appropriate ways to people who interact with them.

But there is not much behind the façade. People sometimes react fleetingly to today's humanoids as though the machines are sentient, but they never really empathise with the robot. Before I die, I want to build a robot that will make me feel guilty, not fleetingly, but reflectively over months or years, if I mistreat it or even slight it socially.

I want to build a robot that I would feel morally bad about switching off.

Rodney Brooks is a leading roboticist. He has written *Flesh and Machines: How Robots Will Change Us* and other books

See an atom

It doesn't get much smaller . . .

Over 10 million of them can squeeze into the period at the end of this sentence. They are the building blocks of matter and the raison d'être of quantum mechanics. Need any more reasons to come face to face with an atom?

For your best chance of seeing an atom with the naked eye, you'll need to make friends with a physicist. Several labs have the equipment you need to trap an atom and cool it until it sits still long enough to watch. You'll have to strip the atom of some of its electrons so that you can hold it in an electric field.

Then, switch off all the lights and bask the trapped atom in green laser light. Barium atoms are a good choice and reflect enough light to trigger the light-sensitive cells at the back of your eye.

It'll take a few minutes for your eyes to adjust to the darkness. Look closely and the atom will appear as a pinprick of light. While you're watching, consider this: the atom you are looking at was forged in the furnace of a star that lived and died long before Earth was even born.

Hear a kakapo booming

The world's biggest parrot is back

If you ever hear a kakapo booming in the mountains of Fiordland in the far south of New Zealand, crack the champagne and celebrate a conservation success story.

Reduced to just 50 a decade ago, the world's largest parrot — a flightless goliath with an extraordinary personality — clawed its way back from extinction with a bumper breeding season in 2002. When summer nights in Fiordland reverberate to a sound like a giant blowing across the top of an outsized beer bottle, it means the birds are back for good. The deep, resonant boom is the sound of the male kakapo trying to attract the attention of females.

Once the birds gathered in gangs and tried to outdo their rivals by puffing themselves up into fat green footballs and letting rip their loudest, deepest boom. The sound travelled across the valleys, drawing lovestruck females from a long way off. Then numbers dropped so low that a whole summer of booming brought no response.

If predictions are right, the number of kakapos living on offshore islands free from predators should grow steadily and eventually there might be enough to let some loose in their old haunts on the mainland. When you hear the sound of the old night bird, you'll know that somewhere a hopeful female is making her way up the mountain.

Home Lab: Make liquid nitrogen ice cream

One scoop or two?

It's the quickest way to amaze and delight unsuspecting guests. And it's the best ice cream you'll ever taste.

For basic vanilla ice cream, mix together two cups of full-cream milk, two cups of thick cream, one cup of sugar, and two teaspoons of vanilla essence in a stainless steel bowl. Then pour in about five cups' worth of liquid nitrogen, while stirring vigorously for several minutes until the mixture has frozen and the nitrogen boiled off. Plop scoops into bowls and top with snazzy-looking wafers. Voilà.

To get hold of liquid nitrogen, you'll need to get on good terms with someone working in a physics or chemistry lab. Nitrogen turns into a gas above –196°C, so you'll need a vacuum flask to keep it cold. And do be careful — a liquid nitrogen burn can be nasty. Wear goggles and protective gloves.

Once they've tasted the dessert, your guests will be begging you to tell them why it's so good. And here's the answer: flash freezing the cream prevents ice crystals from growing too large, while bubbles of nitrogen aerate the mixture, making this the world's smoothest ice cream.

Visit Mars on Earth

It's the next best thing

OK, you'll probably never make it to the Red Planet in your lifetime. But there are a few locations on Earth where you can savour what Mars has to offer. That is if you can find elbow room among the researchers who flock to these Mars lookalikes to test robots, space suits and all the other paraphernalia that would be needed on a manned mission to Mars.

The main requirements of a Mars lookalike are extreme cold and extreme dryness. That's why the dry valleys of Antarctica are the connoisseurs' choice. Not far behind are the Mars Society bases in the polar deserts of Devon Island in the Canadian Arctic, and the volcanic region of Krafla in Iceland.

If you don't fancy the cold, you can always head to the desert near Hanksville in Utah, which is home to another Mars Society base. It may not be the right temperature, but the rocks are red.

The Basics: Read *On the Origin of Species*

Natural selection from the horse's mouth

When you read Charles Darwin's 1859 blockbuster *On the Origin of Species*, one of the greatest minds of all time speaks directly to you. Here, for the first time, is the explanation of how natural selection has created life's great diversity. This is epoch-making science, yet you don't need a PhD or a dictionary of scientific jargon to understand it.

Admittedly, *On the Origin of Species* is no *Da Vinci Code*, but don't let that put you off. Provided you can summon up an interest in pigeon fancying and deal with some stuffy Victorian verbosity, it is eminently readable. And that in itself is astonishing.

Whistle at the northern lights

Make them dance for you

Yellowknife calls itself the northern lights capital of the world, but if you are anywhere north of the 60th parallel and away from city lights, on a clear night there's a good chance you'll see the ethereal curtain of light draped across the sky. Make the journey by dogsled and you will likely hear traditional legends about children dancing in the night, torches held by the dead as they journey to another world, or light shining through holes in the black dome that creates night. Whatever the explanation, even the scientific one seems magical. It's hard to believe that the dance of the green, red and purple light is choreographed by the sun. Every second, one million tonnes of electrically charged material is blown off the sun and hurled across space at more than one million kilometres an hour. In two days the stream of electrons and protons reaches the vicinity of the Earth, where it is funnelled by our planet's magnetic field towards the north and south poles. There the particles are pulled downwards, where they run into atoms of nitrogen and oxygen in our atmosphere, which as a result glow like the gas in a neon tube. While it looks like the lights reach all the way to the ground, they are actually hundreds of kilometres up, higher than space shuttles fly. At first glance the lights seem to be painted against the blackness, but watch for a few moments and they change, colours fading in and out, patches of brightness appearing and disappearing. You are watching a cascade of light, a rain of debris from the sun.

Legend has it that if you whistle or sing, the northern lights will dance for you. Actually, the effect is the vibrations of your head as you make sound, which spread the image over more cells on your retina. But don't whistle too loud or long because those same legends say the spirits can come and either cut your head off or take you away.

Learn an endangered language

The power of the few

In learning an endangered language, you become a repository for those special ways of seeing the world. You become privy to stories handed down through the ages. And you become a vessel of history and of wisdom.

Take Choctaw, a language spoken by Native

Americans in Oklahoma and Mississippi. Today there are only 11,000 speakers of Choctaw. Fewer than 50 per cent of young people are learning the language now, compared with 95 per cent just 20 years ago. Beyond the next two generations, it's hard to know if this language will survive.

Choctaw's unusual feature is that you have to make clear where your evidence is coming from. If you were talking about something you knew for certain was true, you'd use one form of the past tense. But if you were simply passing on information you'd heard from someone else, you'd have to use another.

Choctaw is not the only language to have such markers, but it's a linguistic feature that most of us will never encounter. So the more languages that disappear, the more those special ways of experiencing the world will be lost.

Snorkel on the Great Barrier Reef

Be dazzled by colour

Even with a state-of-the-art TV screen and high-tech photography, David Attenborough's best still can't capture the magic of the Great Barrier Reef. For that, you have to get into the water and swim with the fishes.

In fact you hardly need to swim, just slip in and float mask-down. There, almost within touching distance, is a forest of corals of incredible shapes and amazing sizes. This is the rainforest of the ocean.

You don't have to be in there for long to catch a glimpse of the more spectacular inhabitants. Giant clams gape to reveal shimmering emerald-green lips. Starfish only seem to come in brilliant colours such as scarlet or electric blue. And then there are the fish, all apparently painted by some crazed artist with a penchant for stripes, spots and primary colours.

It's jaw-dropping. Or it would be if your teeth weren't clamped to the snorkel.

Glimpse a sprite

Storm lights

This is the atmosphere's best-kept secret. And to see it, head to a blustery peak at Yucca Ridge, high on the edge of the Colorado Rockies. There you can enjoy a view across 400,000 square kilometres of plains and watch giant thunderstorms raging over Kansas and Nebraska. And if you're lucky, you can spot a luminous sprite dancing above the cloud tops.

Sprites are flickering towers of light that shoot up into the sky above lightning storms. Lasting less than a tenth of a second, these giants flash blue, orange or red and can reach up 30 kilometres. No one is sure exactly how they form, but with its unobstructed views, Yucca Ridge has become a Mecca for sprite-hunters hoping to find out.

Although sprites are usually too faint to see with the naked eye, from here you have a good chance. Look towards a distant storm, shield your eyes from lightning and focus on a region about four times as high as the cloud tops. Now wait.

The most active storms can create one sprite every minute. It will make the hairs on the back of your neck stand on end.

Ride the Bangkok rush hour

Be afraid, the future is coming

You just don't know what rush hour really means until you visit Bangkok. Here nobody's rushing and the streets are gridlocked 24/7. Simmering in your own sweat at temperatures of up to 40°C and 100 per cent humidity, you will be assailed by the stomach-churning stench of rancid, rotting vegetable matter.

If the smog doesn't choke you it will certainly spoil the view, drawing a grey veil over the infinite horizon of nose-to-tail traffic: veteran buses belching out black smoke, brightly coloured tuk tuks, Japanese saloons, motorbike taxis nipping perilously through the jostling vehicles. Throw in the deafening noise of revving, screeching, tooting and honking and it becomes clear that this ride is not for the faint-hearted. This is cutting-edge. It's the future. And the scariest part is that it will probably be coming to a city near you soon.

Watch the night sky

Starry, starry night

Forget the bright lights of Broadway or London's West End, the best show in town is on every night above your head. But you may never know it's there unless there's a power cut because the glow from streetlights veils the sky from city slickers. You'll need to escape civilisation — or at least head into the country — to catch the night sky in all its splendour.

On a clear, dark moonless night, around 3000 stars are visible with the naked eye. For every star visible, there are over 50 million others belonging to the galaxy we call home. Known as the Milky Way, the rest of our galaxy appears as a pale ghostly band arching across the sky.

You can spot other galaxies without a telescope. The southern hemisphere is the place to catch two of the Milky Way's closest neighbours, the Large and Small Magellanic Clouds. And in the north, the Andromeda galaxy lying some 2.9 million light years away is the most distant object you can see.

Plan your visit to coincide with the time Earth passes through the dusty debris of a comet's tail and you'll be treated to a meteor shower. If you're really lucky, you could witness one of the most explosive events in the universe, a supernova.

Touch a tiger

Everyone should touch a tiger. There is no better way to feel the awesome power and feline grace of this most beautiful of the big cats, and to experience its sheer presence. You will then know, in the deepest recesses of your being, that this is an animal that cannot be allowed to disappear from the face of the planet.

Do not attempt this activity without professional assistance. I've done it twice. Once with a 6-month-old tigress who was attached to two burly young men by a strong chain. The other with a gigantic male Siberian tiger, whose sharper parts were restrained by his courageous and devoted keeper on the far side of a heavy mesh fence.

Ian Stewart is a mathematician. He is author of *The Science of Discworld* and many other books

The Basics: Read *A Brief History of Time*

Blow the dust off your copy

It is said that Stephen Hawking's bestseller, *A Brief History of Time*, sits unread on a million bookshelves. One of them is probably yours. So why not blow the dust off your copy, sit down and actually read it all the way through? You might find it less formidable than you feared.

Then you must wait patiently until the book pops up in conversation. Quite casually mention your favourite chapter, your intuitive agreement with the idea of boundaryless time, and your quibbles with the philosophy of Hawking's conclusion.

Don't say that you liked the pretty pictures, though.

Find a meteorite

A planet in the hand . . .

The thought of being the first person to pick up a chunk of rock from an alien world is simply irresistible. You can find meteorites anywhere on Earth, but your best bet is to look in a desert where vegetation and human habitations don't get in the way. Antarctica is particularly good since many of its blue ice fields are littered with rocks that must have originated off-Earth.

Elsewhere, in places such as the Atacama Desert in Chile, you'll need to be able to recognise a true extraterrestrial among the mundane Earth rocks. The likeliest find is an ordinary chondrite that comes from one of the many small chunks that make up the asteroid belt. Look for rocks bearing medium-sized crystals, between which are tiny white spheres called chondrules, which probably started out as molten rock in the earliest days of the solar system.

Another telltale sign is a dark matt coating, called a fusion crust, where the outer part of the meteorite melted as it streaked through the atmosphere. Rarer still would be an iron-rich meteorite chipped off the exposed core of a budding planet. You might even find a piece of the moon or Mars, though that would require an expert eye and a spectacular amount of luck.

Go fossil hunting

What's inside?

There is nothing quite like your very first fossil. So go to a coastal area, such as Lyme Bay in Dorset, pick up almost any rock, and split it open. Inside could be anything from a spiral-shaped ammonite to a fragment of an ichthyosaur, a lizard that swam in the Mesozoic oceans between 250 and 90 million years ago. Be the first living thing to set eyes on it for tens of millions of years.

Dive in a submersible

The underwater twilight zone

If you want to see some really wild wildlife, take a trip in a deep-diving submersible. You could cruise the abyssal plains or plunge to volcanic vents, but for the best under-water safari hover in the twilight zone between 200 and 1000 metres down. Here you'll see the best display of living fireworks that nature has to offer.

The velvety backdrop of the twilight zone is punctuated by flashes, sparks and squirts of bioluminescent light. Ninety per cent of the species here generate light to find a mate, catch something to eat or avoid being eaten.

Switch on a strobe light and the creatures around the sub flash back at you. And when you see that relentless blizzard of lights from inside a submersible, it's almost possible to grasp that the oceans contain a million times as many inhabitants as there are stars in the visible universe.

Home Lab: Meet a tardigrade

They're indestructible

Want to meet the toughest animal on Earth? A beast so resilient that it can withstand pressures six times greater than those at the bottom of the ocean, endure temperatures from 150° C down to absolute zero, and shrug off lethal radiation?

Then gather some moss or lichen from your backyard, wash it and place a few drops of the water under a microscope. If you're lucky, you'll come face to face with nature's most extreme survivor — a tardigrade.

Tardigrades are fascinating to watch. With their cylindrical bodies, up to 1 millimetre long, eight stumpy limbs and tiny claws, tardigrades swim through the water by pawing their legs, hence their nickname "water bear."

They are also one of nature's great enigmas, because

they belong to a largely unknown group of invertebrates that appeared 530 million years ago. Some species reproduce sexually, others asexually. Some species have males and females, others have females only, and many are hermaphrodites.

Their survival secret? Tardigrades can completely shut down their metabolism, surviving without oxygen or food. Of course, any animal can do this but the process is usually called death. Water bears, however, are so tough that they come back to life.

Visit an impact crater

The power of a meteorite

If you want to know how the world might end, take a trip to the imaginatively titled Meteor Crater in northern Arizona. Its changeless looks and desert location mean you can see it for what it is: the most impressive impact crater on the face of the Earth. Wolfe Creek in Western Australia comes a close second.

About 50,000 years ago, a 30,000-tonne meteorite travelling at 65,000 kilometres an hour plunged through the atmosphere and smashed into the ground here. The crater it left is 1.2 kilometres across and almost 170 metres deep, while the rock kicked up by the impact has formed walls rising up to 60 metres. The remains of the meteorite are still buried somewhere deep underground.

Look over the scarred landscape and wonder when the

next impact is due. No one knows. Pray that it isn't the kind of giant that saw off the dinosaurs 65 million years ago. Next time, it could be us.

Decipher the Voynich manuscript

It's a mystery

Find cryptic crosswords too straightforward? Speak seven languages? Bored re-reading *Finnegans Wake*? Then have a go at the Voynich manuscript.

This has a reputation for being the most mysterious manuscript in the world, and it has certainly baffled cryptologists for decades. The apparently medieval text is written in an unknown script, in an unknown language. Its accompanying illustrations of strange plants, star charts and fantastic plumbing hardly help.

It might hold alchemical secrets. Or it might just be gobbledegook. Find out by being the one who decodes it.

Walk with dinosaurs

One minute step for mankind

Standing on a rocky outcrop looking at fossilised dinosaur footprints stretching in front of you is the closest you can get to recreating the time when dinosaurs were living, breathing animals.

Footprints tell you much more about the lives of

dinosaurs than bones: you can learn how big an animal was, how it moved and how it lived. But it's only when you put your foot in the giant imprint left by a three-toed meat-eating theropod, or measure out its stride with your own steps, that you begin to grasp just how big these creatures really were.

Dinosaur tracks have been found all over the world. But your best chance of seeing them is near coasts, quarries, mines and river banks where ancient layers of sedimentary rock have been laid bare.

The tracks in Lark Quarry in Western Queensland were made more than 90 million years ago. Back then the area was a muddy stream leading to a lake. Today, you can see thousands of footprints embedded in shale and sandstone. This was the site of a great dinosaur stampede: over 50 small dinosaurs about the size of ostriches fled when they saw a lone theropod coming. Here you realise that you would have done exactly the same.

Eat the world's weirdest dish

Only the Vikings could stomach this one

These days the best time to eat the world's weirdest dish is during Thorrablot, Iceland's midwinter festival. Known as hákarl, locals eat small pieces of the stuff washed down with copious amounts of schnapps.

You'll need the schnapps when you know how hákarl is prepared. Icelanders first catch the Greenland shark

Somniosus microcephalus. They don't eat it straight away: Greenland shark is poisonous because its flesh contains large amounts of trimethylamine oxide and urea, which stops the shark's blood from freezing but would give you brain damage.

But there is a way to make it edible: bury it underground for six months then hang it out to dry for a few weeks. By then the flesh will be in an advanced state of putrefaction and disgustingly smelly, but the toxins will have gone and you can tuck in.

This rather unusual method of preparation was perfected by the Vikings more than 1000 years ago. No doubt they also liked extra anchovies on their pizzas.

Smell the world's largest flower

It packs a punch . . .

You'll never forget an encounter with the world's strangest plant. For most of its life, *Rafflesia arnoldii* is invisible, consisting of fine threads that grow on vines in the rainforests of Borneo, Sumatra and Java.

Occasionally, where the vines trail over the forest floor, a few fat brown buds burst through the bark like something from the movie *Alien*. Many rot. Some are eaten by animals. Those that survive long enough swell and grow for 9 months before unfolding into the largest, heaviest and most spectacular flowers on Earth, a metre across and weighing up to 10 kilograms.

Stumbling across one of these monster blooms in the wild is startling. But what really gets you is the stench. The five-petalled flower looks and smells like rotting meat, a ploy to deceive flies into pollinating it.

Within days, the whole thing collapses and vanishes. You'll be glad you saw one. Honest.

Paul Davies

Understand nature's mystery number

Numbers have proved fascinating and mystical at least since the time of Pythagoras. Scientists are supposed to resist the lure of numerology, but one number keeps bugging me: it is 137.

The reciprocal of this peculiar number lies at the heart of nature because it describes the strength of electromagnetism. But what is so special about 137?

Its reciprocal arises by combining three basic quantities: the charge of the electron, Planck's constant and the speed of light. Each of these fundamental constants of nature has a value that depends on the units chosen to measure it, but in combination the number 137 is dimensionless — independent of all units. It is a pure number that nature has latched onto. And I want to know, why that number? Why not 56 or 973?

Is this one of those meaningless and absurd things

about the world, one of those "that's just the way it is" things? Or is there some deep and satisfying reason why that number, boring though it may otherwise be, is the number for the job? I've tried all sorts of fancy mathematics to try to derive it from geometry or arithmetic, so far to no avail.

If I ever solve the problem, I'd like to emblazon my coffin with this number, much as physicist Paul Dirac is commemorated by his equation in Westminster Abbey. Dirac tried to derive 1/137 too, without success. His is a hard act to follow.

Paul Davies is a theoretical physicist. His latest book is *About Time: Einstein's Unfinished Revolution*

Experience weightlessness

Fly through the air with the greatest of ease

When you've spent all your life under gravity, weightlessness is a uniquely liberating sensation. The loss of gravity's all-pervading pull means your legs drift up in the air, your arms flail around, and strange things happen to your insides. Veterans of weightlessness can perform impressive mid-air acrobatics, and even novices can practise their forward rolls.

Even if you never make it into space, there are other ways to escape gravity's clutches. For decades, space agencies have carried out research on aircraft affectionately known as vomit comets. These zero-g flights repeatedly soar and dive, leaving you weightless for 25 seconds at a time. It's surprising how long 25 seconds seems when you're bouncing off the ceiling.

To get on board the vomit comet, you'll need to impress NASA or the European Space Agency with a proposal for an experiment in zero-g conditions. Or those with money to burn can buy a place on a Russian flight.

For a more modest fee you can experience similar feelings of weightlessness by taking a dip in a tank of water at a Russian cosmonaut-training site.

Visit Tuvalu

Enjoy it before it sinks

As sea levels rise, this clutch of South Pacific islands is poised to become the first nation to disappear beneath the waves — possibly as early as 2050. Currently located halfway between Brisbane and Honolulu, soon Tuvalu may exist only in cyberspace as the prized internet domain ".tv".

Tuvalu's smallest islands are disappearing fast. Already you've missed the chance to visit the coconut groves of Te Pukasavilivili. Tuvalu's paradise world of coral lagoons, swaying palms and international sex phone lines is doomed because not one speck of land is more than 4 metres above sea level and spring tides regularly take 3 metres of that. Almost every tropical cyclone could bring down the curtain.

So why not board the weekly flight from Fiji to the capital Funafuti while you can? As you travel from one of Tuvalu's four guest houses in its single taxi to its lone bar, speculate on who will own the rights to the .tv domain and the tuna-fishing rights in its vast territorial waters, after the last of its 11,000 Polynesian inhabitants has left. Go now, before it dies.

Win a fortune the mathematical way

Who wants to be a millionaire?

If you're time-rich and cash-poor, then have a go at winning one of the many maths prizes up for grabs. If you're a hotshot mathematician, set your sights on one of seven $1 million prizes offered by the Clay Mathematics Institute in Cambridge, Massachusetts.

To win, you have to prove one of seven conjectures that have given mathematicians sleepless nights for decades. They include the Yang-Mills mass-gap hypothesis, which asserts that if a particle has mass, there is a lower limit to what the mass actually is.

Or you could try hunting for giant prime numbers. Prime numbers, such as 11 or 13, are divisible only by themselves and 1. If you find the first prime with 10 million, 100 million or a billion digits, a US campaign group called the Electronic Frontier Foundation will award you a prize of between $100,000 and $250,000. Your home computer could do all the work. A project at www.mersenne.org offers free software that sifts sets of numbers for primes, although if you hit the jackpot, you'd have to share the bounty with the organisers. Failing that, there are maths prizes starting from $10 at www.mathpuzzle.com.

Or maybe you should just get out more.

Try breath-hold diving

Be a dolphin for a day

For a taste of life as a seal or a dolphin, leave the scuba gear at home and try breath-hold diving. Although we humans can't rival the ease with which our marine cousins move through the water, we do share some of their quirks of physiology.

Immersing your face in cold water, for example, slows your heart to prepare for diving. So take a deep breath and duck beneath the waves, gradually swimming deeper and deeper. Keep reminding yourself to relax and ignore the quietly growing urge to turn back for another breath. It is fed by the gradual build-up of carbon dioxide in the body, rather than dwindling oxygen.

Eventually every breath-hold diver has to acknowledge their air-breathing nature and submit to it. Then it's time to stop swimming and hang suspended in time as your own buoyancy pulls you gently back to the silvery surface.

At the point of turning back, look up. You may be surprised how deep you have gone. The current world record for breath-hold diving without fins is more than 60 metres.

Be patient on the return journey, there is no rush to get there. Expel the acidic gas from your lungs as you broach the surface smoothly. After the freedom of diving like this, you might never want to be encumbered with the paraphernalia of scuba again.

Sleep in a rainforest

Lie back and listen

The Amazon rainforest is a spectacular place, but you've never truly experienced it until you've been there at night. The trick is to string your hammock and mosquito net while it's still daylight. As you're preparing for bed in the darkness, your headlamp will pick out spots of light in pairs all around you — the retinal reflections from myriad waiting spiders. Stop and switch off the lamp for a moment. If you're lucky, the path will glow where fallen leaves are fluorescing as they are gradually devoured by fungi.

Once inside your hammock, reach down below and gather up the skirts of your mosquito net into a tight knot — this is essential if you don't want to be visited in the night by spiders, snakes or other venomous bedfellows. There's no chance of seeing stars through the tightly locked rainforest canopy, though you may spot some fireflies.

But this is the time to lie back and listen to the tree frogs, crickets, eerie nocturnal rustlings, and eventually the screeching parrots and howler monkeys that will greet the dawn.

Anton Zeilinger

Time travel

My dream is to travel back in time, say 200 years, and show people a CD player. How glorious to watch their reactions. Would they be shocked? Would they grab it from me and check if there are miniature people inside? Would some even be creative enough to guess that this is a future technology?

Imagine bringing children from that era back to the present day. I'd like to see their faces when they first saw the lights of aircraft in the night sky. What would they make of the flashing white, red and green lights? Would they be afraid? Would they love it? Would it inspire some of them to be more daring in their lives?

And imagine explaining to those 19th-century children how much longer we live today. And that many of today's 60, 70 and 80-year-olds are healthier, happier, and more active than they are.

It would make me realise just how far we have come and perhaps help me imagine the next 200 years.

Anton Zeilinger is a master of teleporting and quantum computing with light and author of *The Physics of Quantum Information*

Find the first evidence for life

Don't chip this old block

Peering at the earliest signs of life on Earth will take you to Isua in Greenland, a bleak, Mordor-like landscape nudged up against the ice cliffs that mark the edge of Greenland's ice cap. Go in summer when plenty of helicopters are available for charter from nearby Nuuk and the rocks are largely ice-free.

At around 3.8 billion years old, these are the oldest surface rocks on Earth. Mostly they have been baked, mangled, tortured and twisted by the tectonic forces of the intervening years, but one small dark outcrop, scarcely bigger than a saloon car, has mysteriously survived the depredations of time. It's only accessible by helicopter, and to be told its location you will need to promise the scientists who work there not to take even the tiniest chip off their precious block.

Peer at the outcrop closely and you'll find delicate black layers, created as the microscopic bodies of the

Earth's earliest living creatures sank into the sea floor, and which still bear the light carbon isotopes that are the telltale signature of life.

Home Lab: Weigh your head

Take the brainbox challenge

If you've got more brains than brawn, then why not find out just how much your head weighs? Short of having your head cut off and placed on a scale, there's no direct way to do it. But there is a roundabout way: measure the volume of your head instead.

To do this, simply take a bucket of cold water full to the brim, take a deep breath, and lower your head in until the water reaches the base of your chin. Collect the water that spills over the side and pour it into a giant measuring jug — this is the volume of your head.

Knowing this, you can then work out your head's weight by assuming that the brain, like the rest of the body, is mostly water, which has a density of about 1 kilogram per litre at 4°C. So your head's weight in kilograms is simply the same as your head's volume in litres.

Repeat the experiment a few times to get the best reading, or if you're a masochist. A typical adult male head weighs 4.25 kilograms. How does yours measure up?

The Basics: Understand relativity

Tell your friends about Einstein's twins

There's no more satisfying way to master Einstein's special theory of relativity than by being able to explain the puzzling twin paradox. So here's what you need to know.

One twin, Spacey, goes off in a spaceship travelling at nearly the speed of light, while the other, Homer, stays on Earth. Years later, Spacey returns. Relativity predicts something called time dilation, which says that time runs slowly for moving objects. Relative to Homer, Spacey is moving, so time slows down for her. When the twins are reunited, Spacey should be younger.

The paradox comes about when you consider Spacey's point of view. From her perspective, it was Homer who shot off at nearly the speed of light, so Homer should be the younger twin. Both can't be true.

You can resolve the paradox if you understand that time dilation only happens smoothly when considered from a vantage point that is moving at a constant speed. This is Homer's situation. But it isn't the case for Spacey. When Spacey turns around and heads back to Earth, her speed, and with it her vantage point, changes radically. That's the moment when, from her point of view, Homer suddenly appears to age massively. When they're reunited, sure enough, Spacey is the younger one, and the paradox is resolved.

Got it? Great, now try explaining it to a friend in a noisy pub. Test them the next day. If they've got it too, congratulations!

Walk on lava

Masochists feel the heat

Scooping up lava from the large lakes inside volcanic craters is not recommended without protective clothing and excellent life insurance. But certain volcanoes, such as Kilauea on Hawaii's Big Island, have a handy tendency to squirt out ponds of lava from holes in their sides.

You'll need to hike to the pond over very recent lava flows. Although they look black and dead, they are still warm enough to strip the tread from your boots, and if you fall on them their razor edges will shred your skin.

Up close, the heat is like a furnace. You'll need someone both hefty and trustworthy to hang onto you as you lurch over and scoop up some of the molten rock with a geological hammer. Then you can watch this piece of the inner Earth curl and cool into a solid shard of glass.

If you time it right, you can hike back in darkness with the Milky Way overhead and the eerie orange glow of incandescent rock shining through the cracks at your feet.

Answer one of Einstein's questions: Is our universe unique?

I want to discover whether physics is an environmental science. By that, I mean I would love to know the answer to Einstein's famous question: did God have any choice in the creation of the universe?

Unfortunately Einstein used the term God when he meant the laws of physics, but the idea is clear. Are the laws of physics as we know them unique? Is there only one possible consistent universe, or could there be a host of universes with different laws? If the latter is true, then it could be that what we imagine is fundamental is really just an accident of our circumstances. Maybe there are other universes out there in which the electron is heavier than the proton, or there are no electrons or protons.

It would be wonderful to figure out a way to determine the answer.

Lawrence Krauss is a theoretical physicist and author of *The Physics of Star Trek* and *Hiding in the Mirror: The Mysterious Allure of Extra Dimensions*

Stand on top of the world

Enjoy the high life

Forget Everest. To stand on the top of the world, you'll need to head to the equator. The Earth's spin causes our planet to bulge out at the equator by an extra 20 kilometres or so compared with the poles. Even the beaches of equatorial Africa, Indonesia and South America lie farther from the Earth's centre than the summit of Everest. And if you climb to the top of the highest equatorial mountain, Chimborazo volcano in the Ecuadorean Andes, you truly will be standing on top of the world.

If you still have a hankering to stand on the world's tallest mountain rather than its highest point, book a ticket to Hawaii. Mauna Kea on the Big Island rises an astonishing 10,203 metres from the floor of the Pacific Ocean. Measured from base to peak, it's by far the tallest mountain on Earth.

Watch the Earth turn

Follow the shadow

Everyone knows the world goes around, but have you ever seen it move? Take a long stick or use the corner of a building when the sun is low and shadows are long. Go to the very end of a shadow, the point farthest away from the object that is casting it, and place a coin on the ground just to the left edge of the shadow. Watch carefully for only a few minutes and you will see the coin covered by the shadow as the shadow steadily moves from right to left. (The shadow moves in the opposite direction in the southern hemisphere.)

That slow, steady movement, half the speed of a clock's hour hand, is the ponderous turning of the great globe upon which we live. The only other way to see this motion is to watch a sunset or sunrise. Remember, it's not the sun moving, it's the planet rolling. The Earth turns in a counter-clockwise direction when viewed from above, and every day this motion carries you all the way around the planet's axis. A person standing on the equator is moving at 1600 kilometres an hour around the centre of the Earth. Someone in Edmonton is only moving at about 750 kilometres an hour, roughly the speed of a jet airliner.

Have your genome sequenced

Sequence, save, bequeath

There can be no bigger thrill for an egomaniac. After all, what better way is there for friends, relatives and arch-enemies to remember you than to peruse the 3 billion letters of your DNA sequence on a CD or perhaps presented, for those very special people in your life, in a handsome leather-bound encyclopaedia set?

Perhaps this is what Craig Venter had in mind when he sprinkled a heavy dose of his own chromosomes into the pot of DNA that his company, Celera, sequenced as part of their effort to beat the publicly-funded Human Genome Project to the secrets of our genes. Of course, most people could not afford the current $10 million price tag, but in 2002, Venter put the dream within reach when he promised he would try to get the cost down to $1000.

Home Lab: Measure the speed of light with chocolate

It's the cosmic speed limit

Checking out the fundamentals of science is a wonderful feeling in an age when so much makes us feel powerless. And it doesn't get more basic than the speed of light. After all, nothing travels faster than 299,792,458 metres per second. And no matter where you are, or how fast you zip through space, the speed of light is always the same. So why not measure this cosmic speed limit for yourself?

All you need is a bar of chocolate, a ruler and a microwave oven. Microwaves are a type of electromagnetic radiation, so they also travel at light-speed.

Here's how: first remove the turntable in the microwave to allow hotspots to form in the oven. Pop the chocolate in the microwave and heat for about 40 seconds until it starts to melt at the hotspots. Measure the distance between neighbouring globs of melted chocolate. Double this number and you've got the microwave wavelength. You'll also need the microwave frequency, usually printed on the back of the oven or in the instruction manual.

To work out the speed of light, simply multiply the wavelength and the frequency. The best part is you get to eat the chocolate afterwards.

Learn to have multiple orgasms

Oh, don't stop now . . .

It doesn't come any better than an orgasm, so why settle for one dose of bliss at a time? Just think how many more you could fit into a lifetime if you learn how to have multiple orgasms.

It's supposed to be harder for men, but there are plenty of techniques around, with most involving control of the pubococcygeus, or pelvic floor muscle, the one you use to stop peeing. The aim is to prevent ejaculation by squeezing the muscle at just the right time, allowing men to come and come again. Kegel exercises are the first step — work your way up to 30 squeezes, three or four times a week. Sexologists recommend giving your pubococcygeus a workout during tedious meetings, so watch your colleagues closely during the next office borathon.

For determined hedonists, there could be a pharmaceutical fix. According to one small study, cabergoline, a drug used to treat Parkinson's, both enhances men's orgasms and allows them to have several within a few minutes. It is not yet clear if it works for women as well.

The ultimate solution, though, must be the orgasmatron. Currently being developed by a doctor in the US dubbed Dr Pleasure, this spinal implant allows you to have an orgasm at the push of a button. Now where has that remote got to?

Wrestle with a grizzly bear

No pushing, no shoving

Wouldn't it be glorious to stand face to face with one of nature's mightiest, fiercest beasts without fear? Smile at him. Step towards him. Even wrestle with him. And come away unharmed?

Troy Hurtubise, a self-styled inventor and bear researcher based in North Bay, Ontario, can custom-make a grizzly-proof suit for you, for just over $1 million.

Erstwhile models of his allegedly grizzly-proof suit have been rigorously field-tested by Hurtubise. The Ursus Mark VI, for instance, withstood two collisions with a 136-kilogram tree trunk, 18 crashes with a 3-tonne truck driving at 50 kilometres per hour, and strikes by bows and arrows, shotguns and axes. It has yet to be tested against a wild grizzly — but not for want of trying.

The suit might permit more gentle encounters with the beasts too: Hurtubise dreams of filming the birth of a grizzly in the wild and of studying the animals during their long hibernation.

Make blood-red footprints

Turn pristine snow a horrifying shade

For one of life's truly unnerving but oddly beautiful experiences, go to a snow-capped mountain. There, in the melting fields of snow, you may produce a trail of bright red footprints, a disturbing contrast against the pure white of the virginal snow glistening in the sun.

Don't panic! It's not an omen of your impending appointment with the Grim Reaper, nor is it evidence that iron-bearing meteorites have landed, as people used to think. This is your first encounter with red snow — courtesy of a micro-alga called *Chlamydomonas nivalis*.

Ironically, in the brutal cold of this environment the alga is most at risk from the scorching ultraviolet radiation of the sun. To protect itself *C. nivalis* produces a red carotenoid pigment that acts as a sunscreen, filtering out light that could damage its cells, while letting in the wavelengths needed for photosynthesis.

So when you make your footprints, think nature red in tooth and claw.

See Saturn's rings

Halo around an iconic planet

With its rings on display like a peacock in full plumage, your first view of Saturn is a spine-tingling affair. This is more than a planet, this is an icon.

In 1629, the Dutch astronomer Christian Huygens was the first to recognise that Saturn was encircled by rings. With a modest telescope, Saturn springs into view, even under light-polluted city skies. It's not hard to imagine the sense of wonder Huygens must have felt. The faint pinprick about five ring diameters away is Titan, a moon roughly half as big again as ours with a thick atmosphere that astronomers think obscures oily oceans and alien continents.

Titan got even more exciting in early 2005 when a probe named after the astronomer punched through the clouds in search of the secrets below.

Be hypnotised

You are feeling very sleepy . . .

For most of our lives we exist in two different states: awake or asleep. Why not experience a third? Get hypnotised.

Contrary to popular myth, being hypnotised doesn't turn you into a zombie, compelled to obey. Instead, you enter a trance-like, relaxed state where most of the everyday noise gets tuned out. It's like being totally lost in a book or a daydream.

Scientists still don't fully understand hypnosis, and some are sceptical. But it is as if the logical, conscious mind takes a back seat, allowing the subconscious mind —

the part that allows you to do things without thinking — to come to the fore. And it's supposedly very good for giving up bad habits such as smoking or nail biting, though if you take part in any public displays, be prepared for a gentle ribbing when your friends explain how you cradled a shoe in your arms when you thought it was a puppy.

Get hypnotised, and you'll experience a whole new dimension to your brain.

Adam Hart-Davis

Drive through Death Valley

Death Valley would be my destination, and what a place to die. I drove there from Las Vegas in January 1980 and would love to go again in the cool of winter, armed with camera and sketch book. The scenery is truly amazing, from Badwater in the south, with its feathery salt pool, a sign high up on the cliff saying "sea level," and a record summer temperature of 57° C in the shade, to Ubehebe Crater in the north, via the Devil's Golf Course and the rolling sand dunes.

There are two little towns with bars: Stovepipe Wells and Furnace Creek; there are canyons and Joshua trees and rattlesnakes. Best of all, though, is Racetrack Playa, in the far north-west, and so remote there is no building within 50 kilometres. There, on the flat mud surface, the rocks move about in the night. What a place to go.

Adam Hart-Davis is a writer and broadcaster. His latest book is *Talking Science*

See a rocket launch

5-4-3-2-1

If you haven't got enough of the right stuff to become an astronaut, the next best thing is to see a rocket launched into space. It's got everything — spectacle, tension, excitement and noise like you've never heard before.

If it's a shuttle launch, the first thing that will get to you is the sheer size of the craft: it dwarfs everything for kilometres around. And there's always that last-minute doubt. Will the weather be right? Will it be delayed?

If you're lucky, the countdown begins. The tension builds. Then the climax: first a billow of smoke that would put a thunder cloud to shame, then a noise that's like a high-pitched roar, the air around you crackles, and the volume is turned way up. Just as you wonder if you will ever hear again, the rocket lifts off, making a smooth arc through the sky. As it climbs into the blue, the noise subsides and serenity slowly returns.

It lasts only a few minutes, but just recalling it years later will leave you tingling all over.

Watch a total eclipse of the sun

The greatest show off Earth

Nothing can prepare you for a total eclipse. The super-sophistication of centuries falls away when you're faced with total darkness in the middle of the day. It is totally disorienting. After the calls of confused birds suddenly stop, the silence is terrifying. Your senses tingle, the very air around you seems charged. You suddenly understand why primitive peoples found total eclipses so portentous.

Make sure you travel to somewhere within the zone of totality, where the moon completely covers the sun. Seeing a total eclipse from outside this zone is a bit like drinking decaffeinated coffee: you get a taste but you miss the real hit.

If the sky is clear, you can enjoy every facet of the event, from the moon's shadow, racing along at 2000 kilometres an hour, to Bailey's beads, the residual light of the sun passing through lunar valleys. Eventually sunlight will shine out through just a single valley to create the "diamond ring." Then for the real treat: the sun's hot, glowing corona, which is visible only during a total eclipse.

Chase a tornado

It'll blow you away

If you think you've seen it all, then how do you fancy braving thunder, lightning and blinding rain to stalk a kilometre-high column of whirling air that tosses trucks about like toys? Tornado chasing has become an industry in the flat prairie states of the US where warm humid air going north crashes into cooler air heading south. Massive storms called supercells form, and these often give birth to twisters.

Every spring and summer, dozens of teams, professionals and amateurs alike, criss-cross "tornado alley" in the hunt for the perfect storm. To taste the adventure, join a tornado-hunting tour group. First there's the thrill of the chase, deciphering weather maps and looking for clues in the way clouds move. Then there's the desperate race to reach spots where conditions look best.

Get it right and you'll gape as the grey, spinning clouds descend, turning faster and faster until you hear a roar like a jet-powered freight train. You might catch a faint crack as power lines break and trees snap in half. Watch out for flying cows.

Visit an ancient art gallery

Secrets of the caves

It's hard to imagine what prehistoric man or woman was really like, but completely fascinating to try. And what better way than to gaze at the prehistoric paintings created 13,000 years ago during the Upper Palaeolithic period with red and yellow ochre, charcoal and manganese oxide.

Some of the best cave paintings are found around Ariège in the Pyrenees, not far from Toulouse. They were discovered some 150 years ago and serve as a showcase for prehistoric life. Few caves remain open to the public because of the very delicate conditions needed to preserve the paintings. But you can visit the huge caverns at Maz d'Azil, which are as impressive as any cathedral. The rich discoveries in this cave include harpoon heads carved from reindeer and the 12,000-year-old skull of a young girl. And the caves at Niaux include hypnotic drawings of hundreds of animals, including horses and bison.

Walking around this lamplit cave transports you back in time. The paintings on these cold walls give off messages from our prehistoric past.

See a green flash

Blink and you'll miss it

You may have enjoyed loads of great sunsets, but seeing a green flash goes one better. The place to look is the edge of the sun's disc just as the sun rises or sets. If the conditions are right, you'll be treated to a spectacle that few people have witnessed.

You'll need a cloudless sky and, to make things harder, a distant flat horizon — either over open grassland or the ocean. Depending on the atmospheric conditions, light coming from the sun will slow down and be refracted just a fraction. If you are lucky, the refractive index will be spot-on and green wavelengths of light will be refracted the least, making them the last to dip over the horizon, or the first to come up. Keep your eyes peeled, though. If you blink, you'll miss it.

Watch a frog gastrulate

Life as we know it

Forget birth, marriage or death, gastrulation is the most important time in your life. It is one of the real wonders of the world: just how does a human egg know how to turn into a living creature with limbs, organs and head all in the right places?

To see this once-in-a-lifetime spectacular, you need to make friends with a developmental biologist who works on large, easy-to-see frog eggs. Ask to watch some develop. You'll see the fertilised egg cleave first into two cells, then four, then eight, and so on until it forms a hollow ball of many cells. Now watch for the crucial moment: gastrulation.

A slit forms on the hollow ball and cells on the outside spread out and begin to flow though it into the interior of the ball like commuters rushing to work. The embryo now has three distinct layers of cells. These communicate with each other, allowing the nascent body to self-organise before your eyes into a new life.

Build an igloo with an Inuit Elder

Stay in a house of snow and ice

Discover the structural qualities of snow and the secret to survival in the frozen North. An igloo (*iglu* in Inuktitut) may look like a simple arrangement of snow blocks, but if you try to stack blocks in neat circular rows one on top of another, the whole thing collapses in on itself before you finish building it. The traditional snow house, in use by Inuit for the last 10,000 years or so, takes a cue from nature for its strength. A skilled igloo builder arranges the blocks in one continuous spiral from the ground all the way to the top. Through clever cutting and stacking, the snow blocks angle upwards and inwards as they rise so by the time the last piece is put in place, the block is completely horizontal and doesn't fall. One person can make an igloo from blocks cut from within its own circle in about an hour. One half of the floor is cut deeper forming a pit to capture cold air. Once the gaps between the blocks are filled and furs are laid on the upper part of the floor, the interior is transformed into a remarkably warm shelter. A single oil lamp, combined with a little body heat, melts the inner surface of the dome and forms a windproof glaze. This, combined with the insulating properties of snow, traps heat inside so temperatures easily exceed 20°C.

Nomadic Inuit do not claim possession of their snow houses. Once built, they are left for the next hunters to pass through the area. Eventually a network of igloos is left along hunting and fishing routes that everyone shares. Drifting snow forms around the natural round shape of the igloo, making it able to withstand the strongest Arctic winds far better than any nylon tent. And you don't have to carry it on your back.

Be fooled by virtual reality

You can't ignore your senses

If you really want to be tricked by the virtual world, don a headset and head for the famous "Pit" at the University of North Carolina in Chapel Hill. As cameras track your position and feed images into your eyepieces, you walk into a room that appears to have a deep chasm in the middle, surrounded by a narrow gantry. Would you cling to the sides of the room, or ignore what your eyes are telling you and march boldly into the middle? It's much harder than you'd think.

But for the holy grail of virtual trickery you'll need to wait for teleimmersion to come to a lab near you. Forget grainy video conferences. This über-expensive set-up uses some of the world's biggest computers and fastest fibre-optic cables, along with lighting, projectors and even smells to convince you that you are in the room with someone who is really thousands of kilometres away. They haven't managed touch yet, but it's only a matter of time.

Clone your pet

Felix for ever!

If the thought of life without Robbie the Rabbit, Harry the Horse or Mighty Mouse is just impossible, then take heart. One thing you can do before you die — and even after they die — is clone them. Californian company Genetic Savings and Clone now offers to clone cats for $50,000, although their capacity to serve clients in 2004 ran to only four.

But if that's a little too costly even for your favourite furry friends, you can wait for the price to come down. For a mere $295 dollars, GSC will instruct your veterinarian how to remove skin and mouth swabs from your pet up to five days post mortem, and then ship them to the company for storage and later cloning. For now, tissue storage is the only option for dog lovers, since no one has yet managed to copy a canine. Poor old Fido.

Karl Kruszelnicki

Visit the Arctic Circle

My dream relates to the tilt of the Earth. This idea evolved gradually when my family and I crossed 10 of the 12 deserts in Australia over a period of years.

Starting from the equator, we'd drive to the Arctic Circle, probably to Tromsø in Norway. We would do it twice, timing it so that we arrived in Tromsø on the shortest day of the year, and then again, on the longest day of the year.

On our first trip, midday and midnight would appear imperceptible. We would, of course, try to catch the sunrise and sunset outdoors each day, until it became impractical.

Then, six months later, we would arrive in the land of the midnight sun with 24 hours of sunlight. There, I would get a comfortable chair and arm myself with beer, coffee, sunblock and sunglasses. Over the course of 24 hours, I would continually shift my chair so I could keep an eye on the sun that refused obstinately to dip under the horizon.

Oh, and perhaps I might stumble across a proof of mathematician Georg Riemann's hypothesis about prime numbers, which has defied explanation since 1859, and win myself a cool $1 million.

Karl Kruszelnicki is Australia's favourite science writer and broadcaster

---

Measure the Earth with a stick

Think like an ancient Greek mathematician

You can repeat a classic experiment performed more than 2000 years ago by a Greek mathematician who was the first to successfully calculate the true size of our planet. All you need is a friend, two sticks and two cell phones. On a sunny day call your friend in another town that is a known distance away from you. At the same time, go outside and place your sticks in the ground making sure they are standing straight up, and measure the angle of the shadow cast from the top of each stick. The angles should be slightly different depending on how far apart you are because you are over the curve of the Earth. To calculate the circumference of the Earth, take the difference between the two angles, which may be small, and divide that number into 360. So if the angle is 10 degrees, that into 360 gives you 36. Then multiply that number by the distance between the two locations and you should end up with the circumference of the Earth, which is roughly 40,000 kilometres.

This experiment, minus the cell phones, was performed by Eratosthenes, a Greek mathematician who worked in the Great Library of Alexandria in ancient Egypt. His account of the simple experiment is the first record of an accurate measurement of our planet. It must have been a profound moment for Eratosthenes when he made that calculation because the known world at that time was only the Mediterranean, parts of Europe, Africa and India. The Greeks suspected the Earth was round but did they expect it to be that big?

Reproduce asexually

It's me . . . again

When sex is such fun, and still a fairly reliable way of reproducing, why would you want to reproduce asexually? Because some of us just love the thought of producing a clone of ourselves, identical down to the last physical detail.

Worries about dodgy psychology, ethical implications and technical difficulties aside, the Dolly-the-sheep route pioneered by scientists in Scotland in the late 1990s is still one of the most likely in our lifetimes.

But combining repro-tech and clone-tech is also producing some strange possibilities: babies with three parents, two mothers, two fathers, or made of eggs grown from a skin cell.

Other methods for personal Xeroxing are also in the pipeline, albeit a very, very long pipeline. With all the claims being made for nanotechnology, building robots that probe our cells, make copies and assemble them to make a twin should be attainable.

A real outside theoretical method is to use quantum teleportation to copy the quantum states of all your atoms in one location and assemble them in another. More realistically, you could choose a single, representative organic atom of yours to be duplicated. That would be a true carbon copy.

Wear an exoskeleton

Be a superhero for a day

If you've always wanted superhuman strength without working out, popping designer drugs, or coming from the planet Krypton, this one's for you.

There's a special metal suit called an exoskeleton that can transform the weakest wimp. So, strap your arms and legs to the metal frame, connected by mechanical joints. A network of computer-controlled sensors will work flat out to adjust the tension in each joint so you can keep your balance and support the load, while a power pack on your back drives the pneumatic pistons supporting the joints and takes the strain when you lift and carry.

Prototypes are a bit clumsy, but one model built by a company called Sarcos of Salt Lake City, Utah, can already help you carry up to 90 kilograms without feeling the strain. Soon you'll be able to run wearing one, and once you get really used to the extra bulk, you'll find lifting a car or carrying a couple of fridges a total breeze.

Fly in a MiG jet

See the curve of Earth from space

If you want to be the highest human on the planet — for half an hour at any rate — book a ride on a MiG-25 fighter jet. You'll fly all the way up to 25 kilometres, the very brink of space. Look up from that height and the sky is inky black. Look down and there is Earth, curving nicely within the translucent blue hue of the atmosphere. And there's another thrill in store: if you want a break from star-gazing or joyriding at twice the speed of sound, you get to fly the plane yourself.

Several firms have hooked up with the cash-strapped Russian air force to provide these flights. Prices vary, but don't expect much change from a few thousand dollars.

Be levitated

Up and away . . .

How gloriously spooky to lift slowly off the ground, even just a few centimetres. And the good news is that it's not just the stuff of dreams. Back in 1996, researchers in Nijmegen in the Netherlands demonstrated the immense power of high-tech modern magnets by lifting a small frog into the air. The frog's molecules had only just enough magnetism for a 16-tesla magnet to overcome the little creature's weight.

You could ask them for an official visit or, perhaps easier, cultivate your local friendly experimental physicist and get him or her to repeat the Nijmegen experiment in front of you so you can see frogs go a-floating at first hand.

In principle it will work for a human — when someone builds a magnet big and strong enough. Be sure to remove any iron or steel piercings first.

Talk with an elephant

The ultimate trunk call

Watching baby elephants play in Sri Lanka's Pinnawela elephant orphanage is one of the sweetest pleasures the world can offer. But as well as watching the 3- and 4-year-old youngsters splash and trump, think how much more fun it would be to know what they're saying to each other. Or even how the adults manage to communicate over long distances.

Elephants use low-frequency sound to talk to each other and they seem to have little trouble communicating over long distances. But a recent theory about elephants using ground vibrations to talk to each other bit the dust when the team who proposed it later found that vibrations don't travel nearly far enough. So it's probably down to good hearing, even though elephants don't seem to scan for sound.

When it comes to the intricacies of language, researchers in Vienna say they have identified at least 70 distinct elephant signals, using special microphones. One of the young male elephants studied took several months to understand and use the different language elements.

So if you hope one day to take tea with an elephant, there's a chance you may be able to learn enough to pass the time of day.

Fall in love

Head over heels

Even the most unromantic of us can't fail to be blown away by the power of love — that feeling of being totally engrossed, euphoric, unable to concentrate, and insanely happy.

And the funny thing is that it doesn't matter if your belle or beau isn't the most attractive around, your brain will delude you into ignoring that. Buck teeth, spotty face and love handles vanish as a true love potion of mood-altering chemicals and hormones — dopamine, serotonin, oxytocin — turns the world rose-tinted.

The first flush of love feels as good a high as you'll get from any euphoria-inducing drug because it taps into the same brain networks. In fact, it can be hard to tell the brain activity from addiction, and the reward value is as good as gorging on chocolate. How else are you going to experience a little temporary madness? After all, the brain activity and chemicals that keep you obsessing about your new flame are probably the same as those experienced by sufferers of obsessive compulsive disorder . . .

Listen to singing sand

It's an aural thrill

Hearing a beach sing or a sand dune boom long before you can see it is a truly bizarre experience. This isn't just the scratch of grains rubbing against each other, these are glorious reverberations caused by the shear and slippage of sand as fine as flour.

There are over 25 famous singing sand sites scattered around the world and they all sing in different ways. As the sand dunes at Jebel Nabus in the Sinai shift, the overhanging sandstone cliffs focus the sound, making it ring like the rim of a wine glass. The beach on the island of Eigg in Scotland squeals when it is crushed underfoot. Even more bizarre are sand dunes next to an ancient graveyard on Kauai, Hawaii. They bark like a dog, reminding some people of the spirits of the dead.

Why do they sing? It's all down to the shear stress and grain size. A bunch of physicists and engineers have even shown that fine grains of silica gel reproduce the sound of singing sand when they are shaken in a jar. Shame on them for spoiling the romance.

Swim with salmon

Take a trip downstream

Few wonders of the natural world rival the return of wild salmon up the river of their birth and few places host this spectacle as well as the Campbell River on Vancouver Island. Each summer thousands of fish make their way upstream to their spawning grounds. Don a snorkel and wet suit and float downstream while the Chinook, Tyee and Coho salmon swim upstream. A living crimson menagerie passes by your mask as the fish swim around you with as much regard as they would give to a floating log or any other obstacle they have encountered on their long journey.

Kick a buckyball

The tiniest ball in the world

Ever wanted to bend a buckyball like Beckham? Soon you could have your chance. A buckyball is a molecule containing 60 carbon atoms arranged in a structure resembling a soccer ball. Even though it is only a few nanometres across, you'll be able to play football with one when technology devised by IBM in the 1980s gets off the ground.

Back then, the company's researchers envisaged creating a magic glove with pressure pads activated by a gadget called a scanning tunnelling microscope that can sense single atoms. Running the glove over a microscopic terrain of atoms would be just like running your hand over someone's face.

Funding prevented the magic glove from making it off the drawing board. But it would have worked. And, if you could make a magic glove, you could make a magic boot. Imagine putting it on and feeling the hard sphere of a buckyball resting against your instep.

Take DMT

Experience a shamanic state

If your heart is set on a mega-hallucinogenic trip, then some words of warning. First, get a degree in ethnobotany or at least join a proper expedition organised by paid-up anthropologists, botanists, biologists and local guides who know a shaman from a charlatan.

Let a real shaman decide if you're ready to try the grandparent of all hallucinogens, dimethyl tryptamine. DMT comes from the ayahuasca vine in deepest Brazil and it is the stuff that serious psychonauts dream of. Purified DMT is as rare as hen's teeth and illegal in many countries.

But for centuries Amazonian people have been downing a foul-tasting tea made from the ayahuasca vine and drinking it is even a sacramental ritual for two modern religions in Brazil.

So what can you expect? Vomiting to start. Reports of what follows vary from breathing butterflies to moving from a dazzling fast frenzy of colour through full-on hallucinations to another place and territory of mind. And users warn that if you panic, you're in deep trouble. Drink the brew and your experience will last for hours.

Visit the ancestors

Hominids are go!

Long before "Out of Africa" became a mantra for human evolution, anthropologist Louis Leakey was convinced that the African continent was the cradle of mankind. Olduvai Gorge in Tanzania, where he and his wife Mary did much of their pioneering fossil-hunting, remains one of the most awe-inspiring sites for anyone interested in human origins. There, in 1959, the Leakeys made their first major fossil finds with the discovery of the 1.75-million-year-old *Zinjanthropus boisei,* which was later renamed *Australopithecus boisei.*

The gorge, which lies at the edge of the Serengeti Plain near the Ngorongoro crater, was home to our ancestors between about 2 million and 15,000 years ago. Standing at the lookout post among the Masai as they graze their cattle and peering down into the canyon below, it is humbling to imagine early hominids walking through the landscape, and to think how far we've come since then.

See the min-min

Spooked in the outback

The min-min is a fabled apparition of the Australian outback. And to see one on a lonely road in the dead of night is a terrifying experience. It appears as an undulating orb, bathing you in cold whitish light. But what's so scary is how it behaves: bobbing and rushing towards you, sometimes seeming as close as a metre. Even if you career down a dirt track at 120 kilometres per hour, the min-min will follow.

For centuries, the ghostly lights have defied explanation, but now the mystery appears to be solved. So if you're spooked by the min-min, relax: it's all down to atmospheric conditions.

To see one, time your visit to the outback on a night when cold air is trapped near ground level under warm air. If you're lucky, the cooler layer of air will act like an optical fibre, concentrating and guiding light from car headlamps or campfires hundreds of kilometres away.

Even though you know what causes it, the experience will still be a thrilling one.

Stroke a Martian

First contact

There is nothing to beat holding a piece of another planet in the palm of your hand, especially if there is a chance that it contains fossils of ancient life.

The best place to do this is at the Natural History Museum in London, which has a chunk of meteorite ALH84001. Some researchers still insist this bears fossils of ancient Martian life.

It's all very controversial. But even if ALH84001 turns out not to be the real deal, there are wonderful samples of both lunar and Martian meteorites locked up in the museum's subterranean safes, which give us a feel for other worlds. Your best chance of touching them is when they are brought out for a lecture or demonstration. The meteorites from Mars are especially rare compared with ones from the moon, since they have to escape a stronger gravitational field and then travel further through space. Fewer than 20 have been found.

Have millimetre vision

I spy a secret world

Down in the electromagnetic spectrum, beyond the infrared but before microwaves, lies a little-known world. If your eyes could see at this wavelength, the world would be brightly lit by the sun or fluorescent lights. And it would be populated by naked, shiny, metallic-looking humanoids, their skin reflecting the bright sky above. About their persons would hang keys and coins, perhaps even guns and knives, suspended in mid-air as if by magic.

Welcome to the extraordinary, creepy world of millimetre waves. The waves are like a cross between the ordinary light we use for imaging, and radio waves that pass through walls and clothes. They are reflected by anything containing water.

Law enforcement agencies are beginning to wake up to them. They're using millimetre-wave cameras to frisk people from a distance in the hunt for concealed weapons. So take a look the next time you pass through an airport. You could see more than you bargained for.

Write your name in atoms

F.R.E.D.

If you want to make your mark on the atomic world, or leave an unusual legacy, take a trip to IBM's Almaden Research labs in San Jose, California.

There you'll find a machine called a scanning tunnelling microscope that can nudge atoms around a metal surface using a superfine tip. Physicist Don Eigler made headlines around the world 15 years ago when he arranged atoms of xenon on a nickel surface to spell out the word IBM.

Now he's automated his microscope so anyone can indulge in a spot of atom-scale artwork. Visitors to his lab see an atomic surface on a computer screen, which looks like a blurry egg box. To move atoms around the surface, you just point and click on them with a mouse.

M.y. n.a.m.e. i.s . . .

Find happiness

What does it look like?

Why do some people manage to skip through life shrugging off disappointments and setbacks like discarded clothing, while others struggle to survive the daily grind?

The secret lies partly in our genes. These account for around half of the variation in happiness between different people. What's more, good-looking people tend to be more contented, if not downright happy. Granted, there's not a lot you can do about your genes. But don't despair, there's plenty else you can do.

Being sociable, helping other people and having lots of friends all help. And getting married boosts happiness for a couple of years at least.

Then there's money. It certainly won't do any harm, but beware: wealth is a short-term fillip. We quickly adapt our expectations to new-found riches and end up always wanting that little bit more. Envy on the other hand is a sure-fire route to misery.

One last tip: consider moving to Denmark. It's the only industrialised nation where people are happier than they were 30 years ago. Why? The Danes are keeping the answer very close to their chest.

Understand autism

Experience it for yourself

It is perhaps the closest thing we have to an alien world, except on Earth. For some people, noises hurt, smells are overpowering, sudden movements are frightening and even a gentle hug is a nightmare. Not being able to read other people's emotions, refusing to eat runny food because it's too yukky, only being able to learn when someone is sitting next to you rather than face to face. This is just part of what it's like to be autistic in a world created by the non-autistic.

Now researchers are trying to build virtual reality software that simulates features of autism to help others understand the problems faced by the autistic. Very soon everyone interested in other minds could don a virtual reality kit and feel just a little of what it's like to be autistic.

Achieve immortality

We wanna live forever

Woody Allen famously said that the only immortality he was interested in was the immortality achieved by not dying. If you're of the same mind then take heart from the fact that living for ever — or at least a lot longer than we do now — may not be the impossibility it once was.

Researchers have known for a long time that a drastic calorie-restricted diet increases the lifespan of flies, worms and mice by up to 50 per cent. But since stripping away kilograms to add years doesn't sound like a good trade-off, researchers are busy experimenting with resveratrol, a chemical from red wine that seems to mimic the effects of these diets. And some limited genetic tinkering has more than doubled the lifespan of flies and worms, suggesting bigger gains may be on the way.

No one knows if this will work for humans, but volunteers are already subsisting on low-calorie diets. And there'll be no shortage of people willing to sign up for clinical trials of any anti-aging drugs, especially if they're made from distilled red wine.

Home Lab: Mix the world's weirdest cocktail

Now you stir it, now you can't

It has to be one of the most mind-bending experiences you can legally enjoy in your kitchen: manufacture your own dilatant liquid — one whose viscosity increases as soon as you try to stir it, or do pretty much anything else with it.

The easiest way to make it is a simple mix of about 300 grams of cornflour (cornstarch) and 250 millilitres of water. The mixture ripples like water, but instantly solidifies if you dip your finger or a spoon into it. Go slowly and you can put your finger in, but just try pulling it out in a hurry. You can roll the stuff into a ball in your hands, but stop rolling and it just runs through your fingers. Hit it with a hammer and it can even shatter.

Coolest of all, broken-off pieces liquefy and pool together, just like the shape-shifting T-1000 robot in the film *Terminator 2*. Well, almost. But then the T-1000 couldn't thicken sauces for you.

John Sulston

Marvel at a ruler of time

I'd like to visit Shark Bay in Western Australia to see living stromatolites, mushroom-shaped rocks containing intricate textures created by colonies of microbes. Then I'd go inland to see fossil ones. The blue-green algae that make these mounds were among the earliest living things on Earth: the fossils show that they existed 3.5 billion years ago and were responsible for oxygenating Earth's early atmosphere. The fine laminations in the fossil mounds reflect daily growth and may allow the length of the ancient day to be estimated.

Shark Bay would also bring back memories of the time I stood among bristlecone pines in the White Mountains of California, where the method of carbon dating was validated in the 1960s. From the trunks of living trees and their dead ancestors, which lay intact on the ground in these high, dry hills, scientists pieced together a continuous series of rings going back 10,000 years. From each ring, the actual ratio of carbon isotopes for the corresponding year was measured.

Both places act as a ruler of time stretching back into the past and reveal the history of life on our planet. Seeing them, I can marvel at how human thought transcends the here and now.

Nobel prizewinner John Sulston led Britain's publicly-funded effort to unravel the human genome

---

Contemplate the size of the universe

Think big, very big

Why not take a few moments to think about the biggest thing we know of — the whole universe? If you have no idea how big it is, don't worry, neither did Einstein.

So far, studies of distant galaxies have suggested that the universe stretches on as far as telescopes can see. But that might be an illusion. If the universe is relatively small, scientists predict that the space inside it loops around in a bizarre way so that the cosmos has no boundaries. If you were to blast off from Earth in a superfast rocket travelling in a straight line, you might eventually end up back where you started. Galaxies would create odd multiple images of themselves. The universe would be like a hall of mirrors.

As yet, scientists have failed to find the hallmark of this hall of mirrors stamped on radiation left over from the big bang, suggesting our universe isn't small after all. In fact, it's at least 156 billion light years across. Or if you prefer, 1.5 million billion billion kilometres. And that's just a lower limit. Observations don't rule out the possibility that the universe is infinite. Try getting your head round that.

Visit a particle detector

Cathedrals of the universe

Nothing can ever prepare you for just how big physics can be. Your first visit to a giant particle physics experiment will take your breath away. The biggest are at the CERN laboratory in Geneva, Fermilab in Illinois and SLAC in Stanford, California. They are as tall as multi-storey buildings.

Next to each of these experiments will inevitably stand Portakabins housing hundreds of electronic boxes, which process the millions of signals coming from the detectors every second. Go inside and assault your senses: fans roar to stop the electronics overheating, the air conditioning chills you to the marrow, and tens of thousands of blinking lights remind you that particles are being smashed apart inside.

If you can go on a day when physicists are carrying out repairs, you'll be treated to an even greater sight.

Normally the experiment is closed as tight as a clam shell. When it's prised open you'll see it in all its glory: hundreds of detectors, thousands of brightly coloured cables, and even a few physicists dangling from the ceiling on ropes.

Yet as big as this monster is, remember that the paradox of unlocking the secrets of the universe is that you are at the same time delving into the heart of the tiniest things in existence.

See Galileo's finger

Science meets religion in Florence

Galileo's struggles with the Catholic church are famous. His observations of the moons of Jupiter and his support for the Copernican theory on the motion of the Earth led to denunciations and charges of heresy. Eventually, in 1632, the Holy Office in Rome issued a sentence of condemnation. But Galileo has had his revenge in a very Catholic way. In the Museum of the History of Science in Florence there is an awe-inspiring collection of his apparatus, including a telescope made by Galileo himself. And there, looking like a holy relic, is Galileo's own middle finger, preserved in an egg-shaped glass container mounted on a marble plinth. Come here and worship the triumph of reason over dogma.

Be entombed in a pyramid

Let's play pharaohs

If you suffer from claustrophobia, skip ahead quickly. But if you fancy being a dead pharaoh for a few minutes, here's how. A short hop from the Great Pyramids of Giza and the throngs of visitors and hawkers lie the smaller Queen's Pyramids, crumbling and almost camouflaged by sand. At least one has a small room and an empty sarcophagus that you can enter. A guard in a galabiya will approach and switch on the electric light.

Inside, down a narrow stone staircase, is the sarcophagus, and beyond, another set of stone steps that runs up the wall to nowhere. The guard, who'll remain at the tomb entrance, will suggest that you climb these steps for an alternative view. When you reach the top, for a joke (or if you ask him) he'll switch off the light and seal the entrance.

There is no blacker black than the inside of a closed pyramid. For the minute or two that you're alone, your eyes will struggle to focus on nothing, and fail. Alone, in a tomb built to hold a dead queen for eternity, the absolute darkness will get your brain conjuring up whatever nightmare stuff is lurking there. You have been warned.

Dig up Darwin's treasure

Get down and dirty

Down House in Kent, where Charles Darwin lived for 40 years, is a wonderful place to visit. But there's a more personal way to get in touch with the great man. In some of the nearby fields Darwin carried out experiments that led to his other big biological breakthrough. He correctly worked out that earthworms are not a pest, but are vital to soil's fertility.

One experiment involved spreading small stones on the fields and measuring their gradual descent into the soil below — thanks to the slow churning of the earth by millions of earthworms.

Few people know that many of the stones that Darwin placed on the fields are still there, now deeply buried treasure. If you are enough of a detective, you can work out from his book on worms and 19th-century maps just where his experimental fields were. With some careful digging you'll end up with some of Darwin's own stones to inspire you long after you leave Down. Just make sure you get permission to dig first or you'll end up in jail instead.

Make music by playing a temple

Bong . . . bong . . .

If you're the sort of person who loves creating tunes on your cheek with spoons or on the office radiators with a ruler, then this is definitely for you. Among the ruins of the largely abandoned city of Hampi in southern India is the 16th-century Vittala temple complex, and its hall of musical pillars. Each of the 56 ornate pillars is carved out of a single block of granite, and most are designed to sound a note when tapped.

Slender columns cut out of the same block of stone, known as colonettes, help determine the note that each pillar produces. In theory, the pillars mimic a variety of percussion instruments.

Turn up very early in the morning and you might even get this 400-year-old music set, and some of the sounds of the last great Hindu empire, all to yourself.

Have a species named after you

Me-osaurus

What better way to make sure your name lives on forever than have a species named after you? The easiest way to get a namecheck is to befriend one of the world's dwindling number of taxonomists, the people who name most species. If you don't know any, don't worry: send a donation to the Australian Museum's Immortals Program and you might be lucky enough to have a species named after you. But success isn't guaranteed.

To improve your chances, head off into terra incognita and find a species that nobody has seen before. Spend your time looking for plants and insects, because the chances of spotting a novel vertebrate are very slim indeed. Even if you do find a newbie, you're still not guaranteed a namecheck. It's usually the person who describes the species, a taxonomist or other biologist, who chooses the name, so you'll have to convince them you deserve a mention.

Perhaps your best bet is to become a taxonomist. Although it's not the done thing to name a new species after yourself, your name will be listed after the Latin name of any new organism you describe.

And five for afterwards

Boldly go . . .

It's pretty hard to get into space while you're alive, but once you're dead there's nothing stopping you. For a mere $2100, Space Services of Houston, Texas, will have a few grams of your cremated remains loaded on a rocket and blasted into orbit. It's not exactly luxury travel: you'll have to share the rocket with the ashes of other people plus a commercial payload.

But there is the promise of a glorious finale. After a few years, you will re-enter the atmosphere as a shooting star.

Become a stunt driver

If you want to be an unsung hero after your death, you could do an awful lot worse than consenting for your body to be strapped into a car and smashed into a wall. Simply donate your body to medical science and you could be picked to be a stunt cadaver. You could be dropped face-first onto a windscreen. Or impaled by a fast-moving steering column. Safety tests like this have prevented an estimated 8500 road fatalities a year since 1987.

Fertilise plants

Here's how to turn yourself into human compost. Step 1: have your body freeze-dried in liquid nitrogen. Step 2: vibrate gently. Soon your remains will have turned into a fine, dry, odourless power. You can then have them buried in a shallow grave in a small cornstarch coffin, safe in the knowledge that you will quickly turn into excellent compost. Promessa, the Swedish company that developed the technology, recommends planting a nice bush or tree on the burial site.

Nail a murderer

Once you're dead, chances are you're going to lie around decaying anyway. So why not make yourself useful and donate your remains to the Body Farm at the University of Tennessee Medical Center in Knoxville. This remarkable research centre has room for about 20 bodies, which are left to rot in various outdoor locations so that forensic scientists can learn to pin down times and causes of death with ever greater precision. Body farm research has already helped solve more than 100 murders.

Become a diamond

If you don't fancy being compost, think shooting stars are just a flash in the pan, but would still like to become a lasting memorial, this is for you.

LifeGem of Chicago, Illinois, will take a small portion of your cremated remains, extract the carbon, and turn it into a diamond using the high pressure and high temperature of an industrial diamond press. It takes 18 weeks to make a diamond of anything from 0.25 to 1.0 carat, but what's that compared with all eternity?

Plus one thing to do right now

Be in *New Scientist*

Tell *New Scientist* what you'd like to do before you die and your name could appear in print. Your idea might be intriguing, homely, experimental, inspirational or just plain weird. As long as it has a science twist of some sort, send it, along with a few words explaining why you'd love to do it, to:

100 Things to Do Before You Die
New Scientist
Lacon House
84 Theobald's Road
London WC1X 8NS
UK

Or email to 100things@newscientist.com

The best suggestions will be published in the sequel, *New Scientist*'s 100 More Things to Do Before You Die.

Contributors

Thanks to the following for their contributions:

Claire Ainsworth
Alun Anderson
Stephen Battersby
Marcus Chown
David Cohen
Phil Cohen
Jon Copley
Ben Crystall
Kate Douglas
Liz Else
Douglas Fox
Alison George
Mike Holderness
Valerie Jamieson
Graham Lawton
Michael Le Page
Ben Longstaff
Bob McDonald
Maggie McDonald
Lucy Middleton
Alison Motluk
Hazel Muir
Justin Mullins
Sylvia Pagán Westphal
Stephanie Pain
Fred Pearce
Helen Phillips
James Randerson
Eugenie Samuel Reich
Anita Staff
Celia Thomas
Gabrielle Walker
Matt Walker
Christopher Watson
Jeremy Webb
Clare Wilson
Emma Young